# The Islamic Movement: A Systems Approach

KALIM SIDDIQUI

THE OPEN PRESS LIMITED
*in association with*
THE MUSLIM INSTITUTE

London/Toronto
1980

*British Library Cataloguing in Publication Data*
*Siddiqui, Kalim*
*The Islamic Movement — 2nd ed.*
*I. Islamic countries*
*I. Title*
*909 '.09' 7671082 DS35.6*
ISBN 0-905081-06-4

*The Open Press Limited*
*6 Endsleigh Street, London WC1H 0DS*

*Published in the US & Canada by*
*The Open Press (Holdings) Ltd.,*
*338 Hollyberry Trail,*
*Willowdale, Toronto, Canada M2H 2P6*

*Printed in Great Britain by*
*Edwin Packer & Johnson Ltd.*
*Nottingham*

الهجرية ١٤٠٠

1400
HIJRA CENTENARY

## Preface – A New Consciousness

A single world Muslim community – the *Ummah* – and yet so hopelessly divided and fragmented: this is the bleak situation that stares us in the face. It makes the belief in a single *Ummah* appear notional and even romantic. Every practical man, signals the 'scientific' mind, must accept the present divisions as permanent and unalterable. But there are insurmountable difficulties in this otherwise simple expediency of accepting the reality.

The first difficulty is to determine what is fact and what is fiction. For instance, much of modern consciousness of the western-educated Muslim is largely fiction based on untenable assumptions. This consciousness is the creation of a period of history when Islam was no longer the Muslim consciousness. More accurately, perhaps the Islamic consciousness became overlaid by another consciousness.

Thus, if the source of our consciousness lies outside Islam, how can we accept the logic and judgement of such consciousness? Is this not what all recent political movements among the Muslims have been doing – accepting their defeat and dismemberment as permanent? This acceptance of defeat has been taking the form of nationalism and attempts to reconcile with such opposites to Islam as capitalism, democracy, socialism, monarchy, and so on.

At the same time it is also a fact that a very large number of Muslims throughout the world work for Islam and have very deep and abiding commitments. Their short-term and long-term goals are variously stated but in the end amount to much the same. Besides, the unity of the *Ummah* is neither a political slogan put together by an 'International' or a party, nor a man-made theory or philosophy. It does not even depend on a particular interpretation of history. The unity of the *Ummah* is, at its most basic, a gift of Allah. This unity is represented by the

Qur'an and the Sunnah of the Prophet Muhammad (upon whom be peace!). It is, therefore, not open to us to throw away, give up, or renounce as impractical and 'unscientific'. The fault lies with our consciousness, its source, and our intellectual equipment. We have to create a new level of consciousness, re-establish the lost link between our awareness and the Qur'an and the Sunnah, and re-equip our mental faculties. This requires complete revolt against all that prevails today.

Once we free our consciousness from the limitations imposed by the present and rise above the so-called reality we will see the future through the eyes of the Qur'an – we will discover greater unity in diversity than appears on the surface. We will see that the fragmentation is merely at a superficial level; it is merely a function of the schizophrenic state of a handful of the faithful. We will see that the reality that appears so dominant is itself a temporary phenomenon. The true dimensions of Reality are greater than our own time and space. We have to relate our present to the ultimate destiny. We cannot have much faith, and even less imagination, if the height of our aspiration is to make the present 'reality' permanent.

But the transformation of the reality and escape into the future still needs a strategy. One elementary step is to start redefining our own situation and the environment. Our assets that appear small must be written up; the liabilities that appear so large must be written down. The ghost-like figures that keep appearing in the dark tunnel of contemporary history must be challenged and dispelled. In short, we must change our perceptions, expectations and goals.

The paper presented here was written as a first step in search of a new consciousness. If there is a single *Ummah*, then there must also be a single Islamic movement. The force of this logic led me to attempt to discover the unity. At the time when this paper was written, the Islamic Revolution in Iran was some years in the future. The Revolution has confirmed the basic soundness of the approach. Despite centuries of isolation and biases along the Shi'a-Sunni divide, the Revolution's worldview is pitched at the level of the *Ummah*. The Muslim consciousness is rising above the limitations of sectarian belief and geographical and national 'reality' to project the future on the wide screen of the *Ummah*. The new Revolutionary consciousness has also got rid of the schizophrenia of doubt and challenged the validity of the so-called reality. The ghost of the superpowers and the dominant imperialism has been laid.

The unity of the Islamic movement and the unity of the *Ummah* are the realities that will henceforth dominate the Muslim consciousness.

Kalim Siddiqui

*London,*
*The Muslim Institute*
*20 Rajab 1400/3 June 1980*

## THE ISLAMIC MOVEMENT: A SYSTEMS APPROACH

Ideally 'the Islamic movement' should be a behavioural *system* responding to inputs of *information* and changes in the *environment*.

Let us suppose that those of us assembled in Riyadh[1] today from all parts of the world are part of the Islamic movement. How did we come together? There was *information* in Riyadh that we, or our organisations, were 'working for Islam' in our respective countries and in our respective fields. This information in Riyadh was also backed up by the resources of the World Assembly of Muslim Youth (WAMY). A decision was made in Riyadh to convene the Third Session of the General Assembly. The WAMY secretariat communicated this information to us and, having the necessary resources, sent us tickets and visas and made the necessary hotel arrangements. And so we are here!

From this simple model of our coming together in Riyadh a number of deductions can be made.

1. That there is a worldwide 'Islamic movement'.

2. That members of the movement respond to a certain type of infomation in a certain way; in our case the information was the convening of the General Assembly of the World Assembly of Muslim Youth.

3. That the worldwide Islamic movement consists of a large number of small movements or, in the jargon of systems

(1) This paper was first presented at the General Assembly of the World Assembly of Muslim Youth in Riyadh in October 1976 (Shawwal 1396).

analysis, the Islamic movement consists of a large number of subsystems that are to be found in all parts of the world. In size these subsystems range from nation-States, their governments, to small neighbourhood groups around local mosques or *madrassahs*. In between there are international institutions such as the Islamic Secretariat and the Islamic Development Bank; there are also political parties such as the Jama't-e-Islami in Pakistan, great seats of Islamic learning such as Al-Azhar in Egypt, and countrywide youth groups such as the Muslim Youth Movement of South Africa. In range of activity and size the subsystems of the Islamic movement are so many and diffuse that it would be almost impossible to make a comprehensive list of them all or to assemble them at any one time.

4. This diversity of the subsystems of the Islamic movement allows for a wide range of choices for its members. For instance, the WAMY could choose who to invite to Riyadh. And those invited could also choose whether to come to Riyadh. The larger subsystems, such as WAMY, can always secure positive response from a sufficient number of smaller subsystems.

5. This choice at the subsystem level allows each subsystem to choose its own environment. The concept of the environment is important. Technically, everything that lies outside the boundary of a system forms part of the system's environment. But a system might choose to 'ignore' large parts of its environment and 'respond' only to those parts of the environment it either chooses to respond to or cannot avoid responding to. Those of us who are here chose to respond to the WAMY's invitation. We could just as easily have ignored it.

## The Islamic Movement

There is no single organised body or legal entity with a constitution such as a company or a political party with a hierarchical structure which may be referred to as 'the Islamic movement'. Yet the Islamic movement exists and can be assumed to be a *functional* and behavioural system. It has membership, norms, values and goals. A system without structure or hierarchy is a diffuse system. It is also an open system. At its most general level, therefore, it is so diffuse and open that it would, for analytical purposes, be impossible to contain within a manageable framework. Neither is it possible, also from a practical analytical point of view, to define the Islamic movement by membership; to do so one would have to include every 'Muslim' as a member of the 'system' and the 'system' or 'movement' would then be the Ummah as a whole. The Ummah is obviously an entity, and ought to be a worldwide behavioural and operational system. The fact that the Ummah no longer functions as a behavioural, operational system pursuing clearly defined goals, has given rise to the Islamic movement. Indeed, membership might well be one possible way of identifying the Islamic movement. But membership itself presents many difficulties.

For instance, if we include only those individuals who are consciously working towards a goal, we are, unfortunately, excluding the bulk of the Muslim population of the world. For most Muslims throughout the world today the burden of simple survival, or the struggle for the daily bread, is so arduous that, despite the inclination, they have little time and no opportunity to participate in the attainment of the larger collective good of the Ummah. Lest we be misunderstood, let us add also that the Muslims we are thus excluding are not all in the *poorest* section of the Muslim communities. Indeed, and this is the most unfortunate part, the bulk of the 'well off', the 'middle class' and the 'professional' groups must also be excluded; because only a small fraction of Muslims falling in these categories take any conscious part or contribute any part of their income or savings or time to the Islamic movement. They are too busy making money and indulging in conspicuous consumption. Nor can 'education', or lack of it, by itself include or exclude people from the Islamic movement. One of the tragedies of our age is that, through contact with the west and the acquisition of western education, the bulk of the 'educated' might well fall outside the Islamic movement.

One other group we must also, reluctantly, exclude from the Islamic movement. This is the group which would have Muslims withdraw from the worldly affairs (dunya) and concentrate entirely on what they call the 'real life', the hereafter (akhirah), through a method of personal worship which ignores the social order around them.

After the discussion above, who and what have we got left for positive inclusion in the Islamic movement? Fortunately, this process of elimination has still left us with a substantial Islamic movement – small enough to be called a system and large enough to be relevant to the worldwide Ummah. Let us now formulate our definition of the Islamic movement:

> The Islamic movement is a worldwide, open and diffuse system in which individual Muslims or Muslims organised in groups are consciously working towards the reconsolidation of the Ummah into a behavioural, operational and a goal-seeking system.

**Ultra-stable System**

Because of its composition, the Islamic movement is an ultra-stable system. What this means is that the entire Islamic movement can never be destroyed. If an outside enemy attacks one part of the Islamic movement, the enemy can succeed only in destroying some of the sub-systems, even perhaps some of the major subsystems, but never the Islamic movement as a whole. For example, when the British occupied India, they destroyed the Moghul Empire, the Muslim middle class, the Muslim landed aristocracy and the Muslim army of India. The bulk of the Muslims were able to seek refuge in the vast network of smaller subsystems, e.g. mosques and madrassahs scattered throughout India. Ultimately, the surviving parts of the systems that had been destroyed were able to regroup and secure the State of Pakistan when all seemed lost.

Similarly, when Muslim Brotherhood was smashed and its members killed and persecuted, the bulk of them were able to disperse and seek refuge in other brotherly subsystems of the Islamic movement. No enemy of Islam or Muslims can at any one time destroy the entire Islamic movement throughout the world. No number of hydrogen bombs can do the job either. The H-bomb can destroy the United States, the Soviet Union, China, the States of Europe and the Middle East, but nuclear bombs cannot destroy the Islamic movement. Whichever part of the world survives a nuclear war, there will be parts of the Islamic movement among the survivors. In this way the Islamic movement is an ultra-stable system. No power on earth can destroy all of it.

### Behavioural System

The Islamic movement is not a unified 'physical' system in the sense that the biological system of a living organism is a system which can be destroyed by withdrawal of life supports, such as oxygen, food or other essential inputs like water. The Islamic movement is not a purely 'operational' or 'functional' system which, when its operation or function is completed, must either die or perpetrate itself through coercion and the use of power.[1] The Islamic movement as a whole and all its subsystems have (or should have) one common characteristic; *none of them exists for its own sake; the Islamic systems do not have their own selfish motives of survival, expansion, domination and growth; nor do Islamic systems pursue the selfish ends of a group of persons or economically or politically motivated groups and classes or elites.*

The Islamic movement (and all its subsystems) is a 'behavioural system'. The characteristic feature of a behavioural system is that it has goals and norms to regulate its own behaviour (and those of its members) and the ability to resist pressures from its environment. Indeed the behavioural system must be so well organised and adjusted to its environment that the norms and values of its behaviour

---

(1) Systems that become dictatorial or totalitarian are usually those that cannot otherwise justify their continued existence.

become irresistible to others. This was the situation of the Islamic movement for many hundreds of years. The spread of Islam in the early centuries was almost entirely for this reason: the Islamic movement of the time was behaviourally in tune with the norms, values and belief systems of Islam. This caused the Islamic movement of the time to be a dynamic movement, goal achieving movement, liberal and tolerant movement, compassionate and caring movement, God-fearing movement. The behaviour of the Islamic movement and its members was so attractive that most who came in contact with it accepted its behavioural source – ISLAM.

The Islamic movement today remains a 'behavioural system' with Islam as the nominal 'norm' of behaviour. But the decision-making processes and centres have become so diffuse that parts of the system arrive at different decisions and display widely different, indeed contradictory, behavioural patterns. And all this wide range of behaviour is justified under the overall label of Islam. The capitalist has so reinterpreted the prohibition on usury that he can, with clear conscience, go on accepting fixed rates of interest on his loanable funds. The socialist has so reinterpreted Islam that he regards the Qur'an as the 'first book of socialism'[1], and virtually every type of social, economic, political or even personal behaviour is labelled 'Islamic'.

The results of this behavioural diversity in the Islamic movement are only too obvious: the 'Islamic State' is little different in its organisation and 'national-interest' syndrome from a 'non-Islamic State' in today's international relations; many individual Muslims are little different in their mode of behaviour, habits, mannerisms, life-style and pursuits of career and profit from their non-Muslim contemporaries. Obviously, what has happened is that values and behavioural norms of a fundamentally different value system and life-style have found their way into the Islamic movement. Why? and how?

---

(1) For a detailed discussion of this see Kalim Siddiqui, *Towards A New Destiny*, London: The Open Press, 1974.

Let us use the analogy of a physical biological system. A biological system, let us say a human body, if it is healthy and strong, rejects germs that might cause infection. This the healthy body does as a matter of course. Indeed, a healthy and well built body motivates others who might try to adjust their diet and take other measures to improve their own health. But a human body that is weakened by ill-health or old age has a lower resistance level and is more receptive to undesirable infection from the environment. The weak body becomes sick and eventually dies. A behavioural system, however, does not die. Through infection, cross-infection and repeated infection, the behavioural system may undergo such far reaching changes that it may be difficult to identify it with its original behavioural habits.

Such, perhaps, is the present situation of the Islamic movement. It began to weaken (let us date it from the fall of Spain) and has continued to weaken in the subsequent course of history down to today. Simultaneously, there emerged the western civilisation, the nation-State, and the philosophies of nationalism, national-capitalism and national-socialism and communism. The weakened Islamic movement has been infected by all these. This has altered the behavioural pattern of the Islamic movement and all its subsystems.

## Self-correcting System

The self-correcting or the thermo-static quality in physical systems is well understood. This is achieved by providing a system with a feedback loop which enables the system to return to a pre-set 'norm' when it is subjected to environmental change. For instance, the 'auto-pilot' in an aircraft maintains the speed, direction, height and other variables through changing conditions of its flight path. A central heating system switches itself on if the outside temperature drops and switches itself off when the pre-set temperature has been restored. In other words the self-correcting device protects the system from environmental change.

This self-correcting quality in a physical system is a simple matter of design engineering. Nature has also provided the human body and other living organisms similar qualities of adjustment to changes in the climatic or other environmental factors.

In a behavioural system, however, the factor of choice and 'will' distorts the self-correcting function. Take for instance the individual Muslim who spends a whole life-time as a careerist, maximising self-seeking gains by hook or by crook (mostly by crook!) or serving the interests of a 'foreign system'. This individual, when he retires, suddenly takes up the behavioural norms usually expected of pious Muslims – he grows a beard (to conform with the Sunnah!), starts offering prayers five times a day, goes for Hajj, and even makes a few well-publicised donations to Islamic causes! The entire Islamic movement as a behavioural system is roughly in a similar behavioural situation as the old retired individual.

There is, however, one essential difference: the individual dies and at his/her death the role of that individual is over until the Day of Judgement. But the worldwide Islamic movement as a behavioural system, as we noted earlier, is ultra-stable and will not and cannot 'die', nor can it be totally destroyed by its enemies. The nuclear bomb is as impotent as a toy gun when it comes to destroying the Islamic movement, (though parts of it can be severely damaged or even occupied, e.g. Palestine, Lebanon, the Sinai, Kashmir, southern Philippines, Eritrea and many others).

The Islamic movement as a behavioural system has one other advantage over all other comparable behavioural systems; the Islamic movement has the 'norm' clearly defined, the methods of achieving the 'norm' are well known, and the entire system and all its subsystems have a strong and clear memory of the 'norm'. In addition to the 'norm' and its memory, the entire system *expects* that the 'norm' would one day be restored as the system's normal behavioural pattern. The 'norm', of course, is enshrined in revelation as the source of all knowledge

(Al-Qur'an), the method is enshrined in the *Sunnah* of the Prophet (peace be upon him!), and the model was the civilisation of Medina built by the Prophet. Many movements throughout history have been inspired by the ideal of the Prophet's civilisation model and have struggled to bring it about in their modern or contemporary situations. The most recent attempts have been by the Muslim Brotherhood in Egypt and the Jama't-e-Islami in Pakistan. All these movements have, in our jargon of systems analysis, attempted to provide the Islamic movement with a feedback loop which would return the system to its original norm of behaviour.[1]

Indeed, all the subsystems of the worldwide Islamic movement claim, in one way or another, to be 'working for Islam'. We all claim to be 'servants of Islam'. The range and diversity of 'Islamic' work that goes on in every part of the world all the time is the greatest strength of the Islamic movement as a worldwide system. To borrow an old cliché from another field, we can say that the sun never sets on the Islamic movement. Indeed, we may go one step further and prophesy with complete confidence that the sun shall never set on the Islamic movement! This universal quality is the greatest strength of the Islamic movement, and perhaps also one of its weaknesses. Often one part of the system (one or more subsystems) does not know what is happening in another part of the system.

In the Islamic movement as a whole there is a lack of information-creation, information-demand, information-circulation, information-media and information-utilisation. But before proceeding further, let us consider the environment of the system.

---

(1) This has been spectacularly achieved in Iran. The Islamic Revolution there is trying to establish the new system. It is being opposed by many internal counter-revolutionaries who are aided by foreign powers, specially the United States. The Muslim regimes around Iran also appear to be conspiring against the Islamic Revolution there.

## Environment of the System

We have already considered the point that every system exists in an environment. The system is faced with three choices:

1. Try to isolate itself from the environment (i.e. everything that lies outside the system) and thereby become a 'closed system'. But no system can be altogether a closed system, particularly in the days of spy satellites, radiowaves, television, telephone and other means of mass communication. Even a geographically well demarcated system would need to be at least of subcontinental proportions to be able to do without international trade altogether.

2. To try to control as big a part of the environment as possible. (In contemporary international relations, the Big Powers do this successfully and hence the detente over spheres of influence, etc.)

3. To try to adjust to the environmental pressures the best the system can in order to survive. (Small States pursue foreign policies in this framework; they can do little to alter any part of the environment.)

The Islamic movement, however, as defined here, is not a State, though there are some 42 States that are, to varying degrees of commitment, subsystems of the Islamic movement. The contemporary Muslim States – wrongly referred to as 'Islamic' States – are a category, perhaps a special category, but only one category nevertheless, of the worldwide Islamic movement. Defined as it is, the Islamic movement has subsystems everywhere, even within Israel, United States, Western and Eastern Europe, the Soviet Union, communist China, Japan, India or any other part of the world one cares to name. This precludes the Islamic movement from having an 'external' environment in the neat physical sense that a geographically defined State has an environment beyond its boundaries. We must, therefore, look for another way of defining the environment of the Islamic movement.

The difficulty is more apparent than real. In fact by dispensing with the geographically confined 'Islamic world' we have greatly increased the capability potential of the Islamic movement. We are no longer majorities and minorities. It is as important to identify un-Islamic elements or interest groups within Muslim majority areas and States as it is to know the obviously hostile sources. For instance, it would be a mistake to regard the governments of all Muslim States as unequivocally part of the Islamic movement. In so far as governments of Muslim countries are members of the Islamic Secretariat, the Islamic Solidarity Fund and the Islamic Development Bank (and other similar institutions), and in so far as their professions to the cause of Islam are genuine, such governments may be regarded as subsystems of the Islamic movement. But in so far as such governments simultaneously pursue goals of national-capitalism/socialism they are in conflict with the Islamic movement. Thus, hostility, or source of potential conflict or opposition to Islam, is just as likely to be found in Muslim countries as they are to be found in non-Muslim areas of the world.[1] Indeed, in certain circumstances a Muslim government might act in a barbarous manner to destroy or eliminate a subsystem of the Islamic movement. This is perhaps what happened to the Muslim Brotherhood in Egypt.

In terms of the environment, therefore, the subsystems of the worldwide Islamic movement will do well to beware of dangers that might be lurking in unexpected quarters. It is perhaps wise at this moment to say that the environment of the worldwide Islamic movement as a system is the totality of all the behavioural systems that exist anywhere in the world today. In this way we, as members and subsystems of the Islamic movement, shall not take anything, any system, or anyone for granted. Nonetheless each subsystem of the Islamic movement has a local environment, be it in South Africa, India,

---

(1) For a detailed discussion of how successive Pakistani governments professed the cause of Islam but in fact pursued socio-economic policies entirely opposed to the teachings of Islam, see the author's *Conflict, Crisis and War in Pakistan*, London: Macmillan; New York: Praeger, 1972.

Pakistan, Iran, Britain, Libya, Soviet Union, the United States, or anywhere else for that matter.

## A dose of empiricism

Empirically, therefore, the Islamic movement, though real and functioning, is itself not an integrated behavioural system. Its subsystems on the other hand are nearly all fairly integrated behavioural systems within their respective local settings. These subsystems, for the most part, also display high levels of commitment by their individual members. There is also a high propensity to excellence and limited short-term goal attainment. Within limited areas of their operation, the subsystems also display high degrees of competence. All this subsystemic activity, nevertheless, does not add up to output satisfactory in quality or quantity when compared with the output of forces that are opposed to Islam.

Two features appear to be common to all the Islamic work. First, there is a general lack of direction. Individuals and groups (the subsystems) tend to 'work for Islam' in a general open-ended fashion without any clear outcome in view. There is a complete lack of teleological thinking. It is as if there is a silent aceptance that so long as 'something good' is done, it will all in the end add up to Islam! Well, of course, it does not. A lot of unintegrated, un-informed effort adds up to, for the most part, the waste of scarce human, material and spiritual resources of the Islamic movement. Second, most of the Islamic work seems to be undertaken as 'spare time' activity by highly motivated Muslims. Thus, the resources that are committed to Islamic work are, at best, marginal resources. Many leading workers in the Islamic movement see no contradiction in their dual roles in the Islamic system and as pillars of the un-Islamic social, economic and political systems in which they work. This duality of roles seems to have been accepted as an unalterable fact of life. This ambivalence amounts to tacit acceptance, and eventually to active defence of and participation in what should be totally rejected. What begins as a short-term expediency, becomes a habit and the habit becomes a social norm and acquires respectability.

This behavioural schizophrenia among contemporary Muslims is most prevalent in the fields of social, political and economic activity.

The unpleasant fact is that the onslaught of the western civilisation has left few traces of the traditional structures of Muslim societies. The schism between the 'modern' and the 'traditional' sectors of Muslim societies has deepened to a point where neither trusts the other. All the functional political and economic structures and systems in Muslim societies are 'modern', i.e. western and therefore non-Islamic. These structures and systems have their supporters and functionaries. Those who are opposed to such structures and systems and their integration with the worldwide capitalist system, know nothing better. The result also is Muslims' duality of roles in the systemic structures of the contemporary world.

The basic facts would appear to be as follows, that . . .

- While Islam can solve *all* our problems, Muslims can solve *none* of their problems;
- While most Muslims have an *emotional commitment* to Islam, their *operational commitment* is to socio-economic and political structures in the framework of western sponsored and backed nationalist/capitalist/communist forms;
- While Islam demands the total commitment of *all resources*, we offer Islam only our 'spare time' and 'small change' or *marginal resources*;
- While Islam demands that we struggle to replace the existing civilisational forms with a civilisation of Islam, Muslims today are content with, indeed proud of, 'careers' and 'profit maximisation' in alien functional and value systems.

These facts are unpleasant and add up to a grim realisation – that the Islamic movement itself is little more than the conglomeration of Muslims' own marginal resources of all types – human and material. A movement that attracts only marginal resources cannot hope to be more than marginally relevant. Looked at another way, what this amounts to is that the bulk of the resources of Muslims themselves are committed to systems and behaviour patterns that are inimical to Islam. We are opposed to ourselves! The bulk and the best of our own resources are committed against ourselves!

Little wonder then that of all the network of behavioural systems that span the world today, the least effectual is the worldwide Islamic movement and its subsystems.

## The perspective of history

The situation of the Islamic movement described above is depressing. It must, however, be viewed in the perspective of history.

History proper is not a record of events. History is in fact lessons drawn and interpretations made from a selection of facts and events. The historians of the west have so twisted and selected their facts that they have convinced themselves – and succeeded in convincing many Muslims as well – that religion, including their own and Islam in particular, is no longer a sufficient basis for modern, technologically developed, civilisation. This is no place to argue this point, but the absurdity of this view is obvious enough. And it is also obvious why this view has been forcibly deduced from 'history' and even more forcefully foisted as unchallengeable wisdom of the contemporary intellectual climate. The fact is that every new civilisation that manages to become dominant must claim that what existed before was less good, if not downright ignorance. This is the *systemic* and psychological need of the western civilisation.

Let us pause, briefly, and do our own simple sums on history. It is agreed by all that man has lived on earth for many thousands of years. Yet recorded history covers little more than a mere 2000 years. The first few hundred of even these 2000 years are vague, to say the least. The revelation of the Qur'an and the completion of the message of Islam through Prophet Muhammad (peace be upon him!) happened only 1400 years ago. In the first phase of their history Muslims, coming out of the desert, overcame the then technologically developed civilisations of Persia and Rome. They then spread out and established a flourishing civilisation which covered the greater part of the then known world. Islam remained the dominant civilisational force for at least 1000 years before the decline began with the fall of Spain. The last traces of Muslim dominance survived into the twentieth century in Turkey. No other civilisation has a comparable record.

The western civilisation, by contrast, has been dominant for less than 200 years! And already all the signs are that the west has passed the peak; the leading thinkers and philosophers of the west itself are agreed that the beginning of the end of the west has begun. But the problem is that while the west is going down, Muslims are going down with the west, only perhaps a little faster! Yet it is also a fact of history that the only source of an alternative civilisation is Islam. And Muslims alone have the in-depth experience of establishing and maintaining a dynamic, growing and thriving civilisation for far longer than other people in history.

Looking at the world as it is at this precise moment, the failures of democratic-capitalism and centralised-communism (both equally secular and materialistic models) are more obvious than their achievements. Indeed, their inevitable final collapse is writ large on the wall of history. It is only a matter of time.

As human history approaches the next phase of civilisational crises, Muslims must decide what role they wish to play in it. Muslims can decide to remain passive spectators as the west struggles to remain dominant and, if unchallenged, achieves another level of stability for another hundred years or more. On the other hand Muslims can also decide to come out of near oblivion, which is their present condition, and re-establish another civilisation of Islam. The call is, as it was 1400 years ago, to change the course of history and to establish for mankind a way of life ordained by the Creator Himself. The objective is clear: we have to establish a new civilisation of Islam.

## The process and the plan

The Islamic movement is, as we have already discussed, a diffuse worldwide system with a network of subsystems. The Islamic movement is also an ultra-stable system. In its entirety it is not an effective behavioural system. Its self-correcting feedback loops have become clogged with distorting images. On close examination we find that the system's and the subsystem's goal-setting and goal-attaining qualities do not match up to the required capability. Indeed we found that at all levels the Islamic movement gets only 'marginal resources' comitted to it. The main body of Muslims give the bulk of their attention and have the bulk of their resources committed to systems and behaviour patterns inimical to Islam.

Each one of these maladies is primarily because of an environment in which all the information-generation is controlled by sources outside the Islamic movement. The function of information should be to reinforce the value system of the Islamic movement and encourage the continuation of behavioural patterns traditionally approved by Islam. Through the continuous input of information designed to lead to a loss of confidence among Muslims in their traditional behavioural norms, the environment has succeeded in shifting the Muslims' effective operational loyalty away from Islam. And the Muslims who responded to this type of information input and acquired new loyalties to western systems were richly rewarded through a system of economic, social and political patronage. It is these Muslims who, in the main, dominate the socio-economic and political systems in their societies and they also control some parts of the Islamic movement and its many subsystems. Such Muslims, having become self- or class-interest oriented, now commit their marginal resources to the Islamic movement and its various subsystems. And they also manage to feel good about it and even impress others with their 'Islamicity'!

The distortion of images, loyalties and expectations of Muslims has been achieved by the west through a well conceived and brilliantly executed plan to develop the Orientalist tradition of scholarship within the western knowledge industry. This was a fundamental goal which the west had to achieve in order to neutralise the Muslims' commitment to Islam effectively. In this programme the west has invested huge amounts of its intellectual and material resources. The return on this investment has been high. The propaganda type of information, disguised as objective scholarship, has been successfully fed into the mainstream of contemporary Muslim thought.

## Growth goes with knowledge

One other factor should be briefly noted. The Industrial Revolution came at a time when European societies were shaking off their Judo-Christian values and were in any case entering a period of secular materialism. The pursuit of economic growth in a capitalist order unrestrained by moral standards of behaviour led to the creation of unprecedented wealth and power for the European States, especially England. A myth was created that religion had been an obstacle to economic growth. The Industrial Revolution, having occurred in a non-Muslim environment, naturally acquired structures and functions with values and behavioural patterns generally in conflict with the values traditionally upheld in Islamic societies. These economic systems commanded vast resources and provided the engine of expansion for European imperialism. Muslims who felt the onslaught of European imperialism, backed by Europe's industrial and economic power, soon lost the initiative. Those among the Muslims who accepted western life-style and western economic organisational pattern were put in places of power and influence in their societies.[1] European imperialism took care not only to abolish Muslim political authority, but also and more thoroughly to destroy Muslim educational systems, especially the seats of higher learning and research. These were replaced by western-style

---

(1) For a discussion of the Indian Muslims' reaction to European domination, see author's *Conflict, Crisis and War in Pakistan*.

universities and feeder colleges and schools. While the west quite rightly took full advantage of the great advances Muslim civilisations had made in the physical sciences and mathematics, the west proceeded to develop a body of knowledge known today as the 'social sciences'.[1]

There is a fundamental difference between the physical sciences and the social sciences. The laws of the physical sciences, once discovered, are applicable and repeatable by anyone, anywhere and at any time. For instance, once the basic theory of aerodynamics had been formulated and tested by even the most primitive aircraft flying in the air, then the theory could be used for building even better, bigger and faster aeroplanes. But the theory, despite greater articulation, remains the same. The theories of the western social scientists, on the other hand, are, by and large, *post facto* rationalisations of social realities that developed in the wake of social changes that were as unplanned as they were unexpected. The whole of the economic theory that now supports, explains and justifies the capitalist system has been written to rationalise a system that had developed without the help of and entirely free from any value systems or behavioural norms. Subsequently Karl Marx and John Maynard Keynes did pioneering works. Marx's theories, when applied, had to be almost entirely rewritten by Lenin and later by Mao in China. Keynesian capitalism gave the capitalist system a relatively trouble free 25-year period of high growth rates and full employment.

The fact that concerns us here most is that all the social sciences of the west reflect social orders and have no relationship or relevance to Muslims, and even less to Islam. If we learn and apply western social sciences, then we are not serious about Islam.

---

(1) The west's traditional wisdom about the 'nature of man' and man's alleged propensity to conflict behaviour is discussed in the author's book, *The Functions of International Conflict*, Karachi: The Royal Book Company, 1975

The first priority, therefore, must be the development of integrated academic disciplines of economics, politics and sociology and alternative operational models for a future civilisation of Islam. The absence of these disciplines in the epistemological framework of Islam is the primary cause of the wide and still widening gap between the Muslims' emotional commitment to Islam and their rational, operational, and actual commitment to socio-economic and political systems of alien origin. Belief based conviction in the supremacy of the teachings of Islam does not carry with it the behaviour that follows because belief is not backed up by rational conviction. So long as this intellectual stagnation lasts, Muslims will go on committing only their marginal resources to Islam.

The worldwide Islamic movement, together with its sub-systems, needs the cement of a new intellectual tradition of thought, operational plans, and short and long term civilisational goals. Once this framework has been developed it will form an information network binding the Islamic movement into a coherent goal-oriented combination of forces that will then become unchallengeable and unbeatable. It will so infuse the Muslim mind, spirit and body that Muslims will begin to commit the main body of their resources to bring about the civilisational forms of Islam into a real, living, dynamic, growing, and thriving social system.

Until this happens, the world will continue to ignore us. They will not, and who can blame them, take us seriously. While Islam will continue to remain capable of solving all our problems, Muslims will also continue to be quite incapable of solving any of their problems. The Islamic movement will remain a 'fringe' activity while history will continue to be shaped by the forces of evil.